VISUAL SELLING AND DESIGN

VISUAL SELLING AND DESIGN

JUDITH OWENS, Editor
Vice President Sales Promotion
Retail Service Division

MARY IRISH, Writer

NATIONAL RETAIL FEDERATION INC.
Retail Services Division

Distributor to the book trade in the United States and Canada:
Rizzoli International Publications, Inc.
300 Park Avenue South
New York, NY 10010

Distributor to the art trade in the United States:
Letraset USA
40 Eisenhower Drive
Paramus, NJ 07652

Distributor to the art trade in Canada:
Letraset Canada Limited
555 Alden Road
Markham, Ontario L3R 3L5, Canada

Distributed throughout the rest of the world by:
Hearst Books International
105 Madison Avenue
New York, NY 10016

Copyright © 1990 by
PBC INTERNATIONAL, INC.
All rights reserved. No part of this book may be reproduced
in any form whatsoever without written permission of the
copyright owner, PBC INTERNATIONAL, INC.,
One School Street, Glen Cove, NY 11542.

Library of Congress Cataloging-in-Publication Data

Irish, Mary.
   Visual selling and design / by Mary Irish.
     p.   cm.
   ISBN 0-86636-122-7
   1. Show-windows—Pictorial works.   2. Display of merchandise-
—Pictorial works.   3. Shop fronts—Pictorial works.   I. Title.
HF5845.I77   1990
659.1'57—dc20   90-32639
                                                                                        CIP

CAVEAT—Information in this text is believed accurate, and
will pose no problem for the student or casual reader.
However, the authors were often constrained by information
contained in signed release forms, information that could
have been in error or not included at all. Any
misinformation (or lack or information) is the result of failure
in these attestations. The authors have done whatever is
possible to insure accuracy.

Color separation, printing, and binding by
Toppan Printing Co. (H.K.) Ltd. Hong Kong

Contents

FOREWORD   *8*

INTRODUCTION   *10*

CHAPTER **I**
Shakespeare Might Have Made a Great CEO   *12*

CHAPTER **II**
Sight Lines and Good Designs Direct Traffic   *30*

CHAPTER **III**
One Buzzword of the '80s Survives and Thrives   *40*

CHAPTER **IV**
The Merchandise is the Message   *50*

CHAPTER **V**
**Wit: Dangerous But
Delicious** *64*

CHAPTER **VI**
**Graphics, GRAPHICS, *graphics*,
graphics** *82*

CHAPTER **VII**
**Great Spaces: An Endangered
Species?** *102*

CHAPTER **VIII**
**Small Space: A Special Sort of
Charm** *110*

CHAPTER **IX**
**Christmas Glories and Other
Gala Events** *122*

INDEX *154*

FOREWORD

CHOICES

With over 35,000 retail shopping centers across the country, Americans have a myriad of choices. And, we are more visually sophisticated and better informed than ever before to make these choices. We want the options to be exceedingly clear. We do not have the time, or the inclination, to shop around like we used to. We are working at work, working at play; we do not want to work at shopping.

The American customer is beginning to sound a bit like the character Howard Peal in the movie *Network News*: "We're mad as hell and we're not going to take it anymore!" We are deserting our traditional shopping grounds and beating a path to the stores that have a personality we can relate to; stores that appeal to our lifestyles, values and aspirations — places where shopping is a pleasurable experience, where we are valued as customers, and where the time and money we spend is appreciated.

How do customers make choices? From the first glimpse of the store front to the final touches in the fitting room, people respond to color, design, lighting, texture and details that articulate a store's personality and image. People choose based on their perception of who you are, what you sell and how you sell it.

Our job as professionals is to create a three-dimensional environment that defines a store's personality and presents the merchandise to the customer. The better our choices in store planning and design and visual merchandising, the easier it is for the customer to choose.

Depending on a store's image and positioning, its retail environment can be fantasy or function, buttoned down or bee bop; whatever expresses the appropriate personality and style. People respond to style, whether it is a multi-store mass merchandiser or a one-off mom & pop style that transcends price. It draws customers through the door and brings them back for more.

While a store's personality and style are important, the currency of the day seems to be Time. Precious to us all, there just doesn't seem to be enough of it. Between information overload and work and family responsibilities, the average American enjoys approximately 40 percent less leisure time than he/she did fifteen years ago. Is it any wonder we get impatient with stores that do not make shopping easier?

In addition to creating the right imagery and telegraphic merchandising messages, we can also help customers to help themselves through logical and thoughtful planning. The best visual merchandising in the world is useless if you cannot get people circulating through the store to see it.

Developing the right equation and balancing the planning, design and visual merchandising requires making choices — sometimes tough choices. There are (almost always) risks. But, when the choices are made based on a focused and well-positioned image, and on knowing your customers' needs and expectations, there are (almost always) rewards.

VISUAL SELLING & DESIGN is a celebration of the art and science of balancing that equation. Bravo to all the professionals whose work is presented in this book. You took risks. You made choices that made it easier for your customers to choose.

Paul B. Humes
Principal
NBBJ/Retail Concepts

INTRODUCTION

Every customer we seek to entice into our stores is bombarded from every which way with visual clutter. Television, magazines, catalogs by the hundreds, billboards, and transit advertising. Store, mall and supermarket environments are awash with colors, neon, signs, and products. Even the cars our customers drive and the clothes they wear are emblazoned with rhetoric.

That puts heavy responsibility on the Visual Merchandiser to cut through visual clutter. To simplify. To beguile and befriend real people with real needs and pressing time shortages. To become sophisticated marketers stretching to streamline the visual tools of the visual merchandising trade. To SELL!

From the NRF's catbird seat, we see power in the '90s accruing to creators of visual selling and design tools which reach out to make your store the choice of your customers.

In the pages of this book, you'll see examples—some simple, some small, some grand, some glorious, some graphic—of visual merchandising and design that sells. Enjoy!

John J. Schultz

John J. Schultz
President
National Retail Federation Inc.
Retail Services Division

CHAPTER 1

Shakespeare Might Have Made a Great CEO

We all know the retail industry's favorite Shakespearean quote: "Costly thy habit as thy purse can buy." But here's another that some of America's most successful merchants have adopted: "To thine own self be true." They've evaluated their markets, chosen a niche and capitalized on one particular "personality" — their own.

A store's persona may be inherited, as with ZCMI; refurbished and refined, as with Bergdorf Goodman; or created from scratch, as with Esprit. It may be undeviatingly upscale, young and free-wheeling or middle of the road family-oriented and still be fine, famous and financially healthy. One vital key is the courage to be consistent, to resist the temptations of a quick fix — an ever changing image. Wise planners know that in these days of specialization, casting a wide net seldom works. A busy customer (and that means every customer) who's confused about what sort of store you are, won't waste time wondering and wandering your aisles.

On the following pages are examples from four stores, dramatically different from each other, which have used visual merchandising clearly, coherently and above all consistently, to communicate their very individual images. See what happens when creative professionals apply integrated design to illuminate the adage, "Be Yourself."

ZCMI fashion news, not haute couture, but fresh and interesting thanks to Tim Taylor, designer.

Even a Martian landing in Salt Lake City, Utah, would no doubt feel welcome at ZCMI. Mike Stephens, Director of Visual Merchandising, makes sure the store has an aura of warmth and friendliness beginning with the windows. Graphics with crisp copy are one key. Bright, high-focus lighting is another. Simple foam-core columns add drama without detracting from the merchandise. Thella Hall, designer.

More good ZCMI graphics. Don't miss the wit of the over-sized light cord reinforcing the copy line. Sue Hansen, designer.

A slightly surreal spring in Salt Lake City. ZCMI designer Roy Lauritzen uses silk screened topiary above a "floating" floor. Notice how your eye is drawn to the merchandise.

Taking advantage of architecture to make a focal point that's fun and fascinating. The Esprit Superstore in Georgetown, Washington, D.C., by the Esprit Georgetown Visual Display Team.

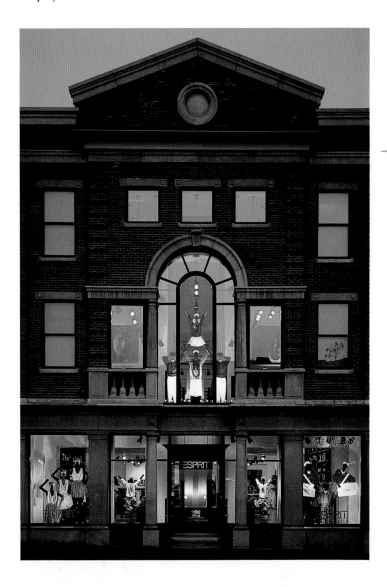

Sharp lighting and a lot of negative space creates a gallery-like atmosphere. See how the cool simplicity of the setting contrasts with the warmth and intricacy of the sweater designs. A wholesale showroom presentation by Esprit's San Francisco, Ca., Display Team.

Never a doubt about Esprit. All the world knows it's young, confident, casual. It's consistent even when it is a shop within a store, like this example from Macy's California. Esprit "trademarks" here include industrial-look architecture in a simple, open environment, logo-lettered graphics, and a photo blow-up that pinpoints their customer profile. Both display and merchandising by Esprit's San Francisco Display Team.

ZCMI designer Sue Hansen shows us how to build a summer wardrobe: hot colors and lots of black. It's more fun minus mannequins. Smart graphics tell the story.

The same elements in another Esprit shop-in-a-shop. This time the junior area at Macy's California, with display and merchandising by Esprit's San Francisco Display Team.

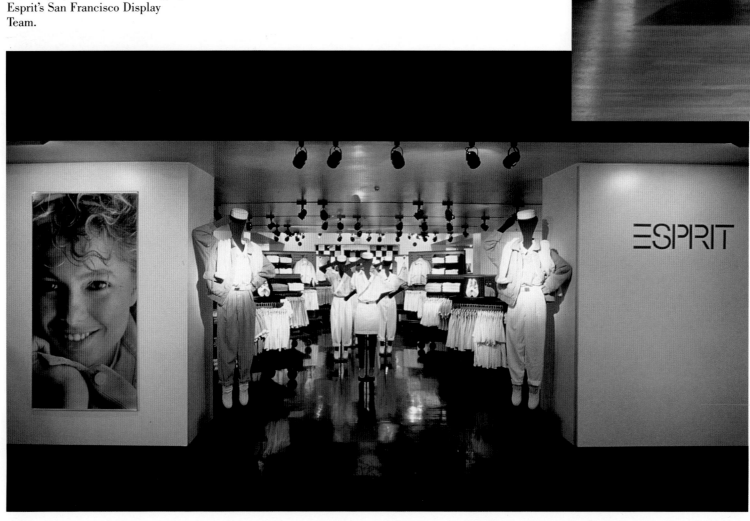

Even the color is consistent here. Esprit's Dallas, Texas, Visual Display Team utilizes all-white for the Esprit Superstore in Dallas.

Flowers, repeated on screens and chairs, and a chandelier light this romantic lingerie vignette. William Short re-emphasizes tradition at Parisian's Forest Fair Mall store.

The entrance to Parisian at Forest Fair Mall in Cincinnati, Ohio, uses graphics, sleek space and abundant merchandise to communicate the store's personality instantly: updated, sophisticated, yet traditional. William F. Short, Director of Visual Merchandising; store design by Schafer Associates.

Following through on the romantic theme at Parisian, Forest Fair Mall: antique-look doorway designed by Jim Mitchell, Store Planning. Antique props are used throughout the store to keep the traditional feeling consistent.

The Empire theme is carried into a home furnishings display. By Angela Patterson and Richard Currier for Bergdorf Goodman.

While the rest of the world was celebrating the anniversary of the French Revolution, Bergdorf Goodman, New York, was one step ahead, into the Napoleanic Empire — a subtle way to say "we're avant garde." Angela Patterson, Vice President, Director of Store Planning and Design, and Richard Currier, Director of Visual Presentation, used huge, exquisitely lit paintings (richer looking than photographs) to stop Fifth Avenue traffic and to make the merchandise seem discreetly elegant by contrast.

Another subtle idea here. Is that George Sand, Chopin's notorious lady-love, that the mannequin is admiring? Further touch of discretion: the quiet graphics that announce Chanel Couture. Also by Angela Patterson and Richard Currier of Bergdorf Goodman, New York.

More glories of the empire. There is continuity in the use of grand scale art and the Empire theme, and consistency in the message: this is clearly a very sophisticated place. Window by Angela Patterson and Richard Currier for Bergdorf Goodman, New York.

CHAPTER II

Sight Lines and Good Designs Direct Traffic

Retailers will recognize these problems. You've practiced what Chapter I preached, Consistency. You've captured a loyal customer who likes you just the way you are *but* has a habit of making a bee line to your Bath Shop, collecting some loss-leader towels and leaving the store. Then there's the bitter buyer who bewails her shop location: "How can I ever make my figures with my merchandise stuck off in that corner?"

Store designers and specialists in visual merchandising will be interested in these examples of techniqes their counterparts have used to solve problems. How to turn that towels-only consumer into a lady who says "I had no intention of buying evening pumps today. I just somehow found myself in the shoe department." And how to switch that buyer's sulks to smiles.

Everyone who has ever shopped will be fascinated and flattered to discover that psychology, applied fine art and a lot of careful planning have gone into what may seem at first glance like just another pleasant atmosphere.

   Look these photographs over carefully and you'll see that each setting was created to control, direct and influence traffic within the store.

Bloomingdale's New York built an in-house highway for its fast paced California promotion, and used a lane divider as a clever sight line to drive us right into center-store action. Attention-getter here: mannequins and their motor bikes. By Joe Feczko, Vice President of Visual Presentation, and Robin Lauritano, Creative Director of Visual Presentation.

This must be the place; the sign post tells us so. A traffic directing device is also a destination. By Joe Feczko and Robin Lauritano.

Sight lines under foot, lighting sight lines over head and a sign post show the direction. By Joe Feczko, Vice President of Visual Presentation, and Robin Lauritano, Creative Director of Visual Presentation.

What buyer wouldn't be delighted to preside over a shop like this? And what customer could resist entering? The understated silver lettering, right, tells us it's Yves Saint Laurent's headquarters at Bergdorf Goodman, New York. By Angela Patterson and Richard Currier.

Another "destination" location at Bergdorf Goodman. The Christian Lacroix shop has its own ambience, completely different from YSL. And don't miss those fixtures with C.L. hangers, a subtle reaffirmation of the designer's status. By Angela Patterson and Richard Currier.

Who could resist strolling to the very end of this festive, flag-draped aisle, with twinkling lights beckoning at its horizon? Thomas Azzarello, Vice President of Visual Merchandising at Emporium-Capwell in San Francisco, Ca., took advantage of the store's great space to create a gala Christmas welcome at the main entrance.

Horizontal ceiling lighting becomes a magnet which pulls us into the junior department. Columns form an inviting open doorway with neon capitals placed strategically under two square ceiling lights. By William Short for the Parisian store at Forest Fair Mall.

The same aisle, dressed for spring, draws customers down what seems like sunny miles. By Emporium-Capwell's Divisional Vice President of Visual Merchandising, James Bellante.

Seeing double. Delightful double dots and the extra surprise of a male mannequin catch a browser's attention, while the traditional pine curiosity cupboard draws her further into the better ready-to-wear area at Parisian in Forest Fair Mall, Cincinnati, Ohio. William F. Short, Director of Visual Merchandising.

A soft focus accents the designer salon of Bonwit Teller in New York. In this case the shopper is drawn toward the daylight appeal of glass and the inviting curves circling within curves. By F.C. Calise, Vice President, Visual Merchandising and Design.

This irresistible, easy little stairway lures the customer up to an "inner circle," spotlighted and softly lit from above. The angles of the stair rail add contrast and mystery to the roundness of the area above. By F.C. Calise for Bonwit Teller in Cincinnati.

CHAPTER III

One Buzzword of the '80s Survives and Thrives

"**L**ifestyle." Ah, if each of us had just one dollar for every time we've heard it, read it, said it. Yet it's still valid. Lifestyle is the clearest way to describe one of the best tools visual merchandisers can use. Here's how it works: Create a special corner or atmosphere, and watch the customer respond.

First, there's *recognition*. The shopper looks at a setting and feels at home in your store. "These people understand me. This is how I am (how I dress, how I furnish my home, how I live). I know I'll find what I need here."

Secondly, *aspiration*. "This is how I'd like to be, look, live." Here a designer is also an instructor, using everything from well-dressed mannequins to beautifully set dining tables to show upwardly mobile customers how to change, improve, move ahead in the world. Of course, it's a way to sell; it's also an exciting way to serve your store's clientele.

Be aware of some new catch phrases for this new decade. "Life Stage" is a reference to a particular part of a customer's life cycle. He or she may be setting up a first apartment or bringing up a teen-ager or be a member of the generation that is gaining growing interest among market researchers: the affluent over-fifties who often lead more active social lives in smaller, smarter homes. "Mind Stage" presumably means that what you think is of more importance than your age.

In the meantime, study these examples of displays keyed to diverse lifestyles.

When Bloomingdale's, New York, decided to do a California promotion, they didn't just stay in the muscle beach mood. They divided the state into 12 Californias, each with a lifestyle of its own. This bedroom has the relaxed but richly traditional feeling typical of northern California. Designed by Richard Knapple.

If the L.A. motorcycle brigade looked this chic, wouldn't you like to join them? By Joe Feczko, Vice President of Visual Presentation, and Robin Lauritano, Creative Director of Visual Presentation for Bloomingdale's New York.

Life in a California kitchen: bright, light and full of fun. While Esprit eschews props, Bloomingdale's New York goes all out, with everything from plants to copper pots. By Joe Feczko, Vice President of Visual Presentation, and David Hamond, Home Furnishings/Designer, Visual Presentation.

A totally different sort of bedroom, minimal and modern, is set up inside a sleek, simple kiosk complete with a blow-up photo to identify with the customer. A wholesale showroom by the Esprit San Francisco, Ca. Display Team.

This kiosk gives us a quick and easy insight into the Esprit Man. We don't even need a mannequin to see that the lifestyle is casual and includes plenty of pretty women. By the Esprit San Francisco Display Team, for an Esprit wholesale showroom in San Francisco, Ca.

Ask any young urban professional if she'd like to look like these mannequins-on-the-rise and she'll answer, "YUP!" Window at the Esprit Superstore in Los Angeles, Ca., by the Esprit Los Angeles Display Team.

The Southwest lifestyle at home. Painted bed, Indian-look blankets and correct accessories, including a conversation-piece skull. By Joanna Seitz and Robert Whittingham for J. Seitz, New Preston, Ct.

The Southwest, very much in fashion, comes to life in the window at Henri Bendel, New York. Danuta Ryder, Visual Planning Director, recreates the country lifestyle, complete with blue skies, cacti and real sand under foot.

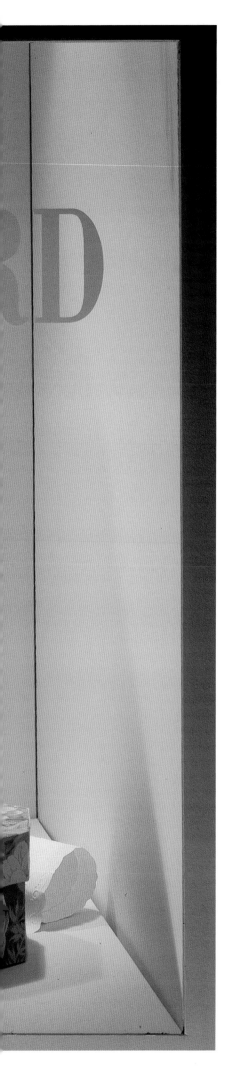

Now this is the life every mother should lead: feminine, flowery and luxurious. The merchandise itself makes perfect props. Note how the "young" copy and graphics contrast with, yet complement, the classic pastel backgrounds. Mike Stephens, Director of Visual Merchandising, Thella Hall, Designer. ZCMI, Salt Lake City, Utah.

CHAPTER **IV**

The Merchandise is the Message

This, after all, is what every store, Fifth Avenue fancy or no-frills friendly, is about: buying goods and selling them at a fair profit. The merchandise is the heart of the matter and no wise store designer buries it. Beware the temptations of too many props, no matter how clever, and forego the too too tres chic salon with chaise longue, lacquer table and no clothes.

A good merchandise display moves goods, fast and frequently. But it does much more. Chapter III touched upon a point that may well be the most important one for retailers in the 90's, Customer Service. Stores that offer what is perceived as service win customers. Patrons no longer have that "my store right or wrong" loyalty; today they ask, "what have you done for me lately?"

Teaching, by design and display, how to decorate a den or accessorize a dress is a service you can offer without extensive expenditures, a service customers appreciate and enjoy. Do it consistently and you'll earn an extra bonus, a reputation for authority and leadership in fashion and/or home fashion.

No doubt you'll recognize some of these photographs from previous chapters, which proves that well–planned displays can do double, even triple duty. You'll not likely need anyone to explain these messages. They speak for themselves, as they should.

By Joanna Seitz and Robert Whittingham for J. Seitz and Co., New Preston, Ct.

This tempting display shows shoppers how to accessorize with a Southwestern flair. By Joanna Seitz and Robert Whittingham for J. Seitz and Co., New Preston, Ct.

By Joanna Seitz and Robert Whittingham for J. Seitz and Co., New Preston, Ct.

When classic white shirts are back and red repp stripes are in, don't be shy about it, says Tom Azzarello. He emphasizes the news not once but eight times at the Emporium-Capwell in San Francisco.

The world is mad about plaid. Danuta Ryder, Director of Visual Planning, announces the news in the window at Henri Bendel, New York.

Danuta Ryder of Henri Bendel makes glamour a mirror-clear case of black and white.

Call them suspenders or call them braces, they make a double fashion impact when shown with bow ties. By William Short, Director of Visual Merchandising, for the Parisian at Forest Fair Mall, Cincinnati, Ohio.

The merchandise is the message, even in this designer salon at Bergdorf Goodman, New York. By Angela Patterson, V.P., Director of Store Planning and Design, and Richard Currier, Director of Visual Presentation.

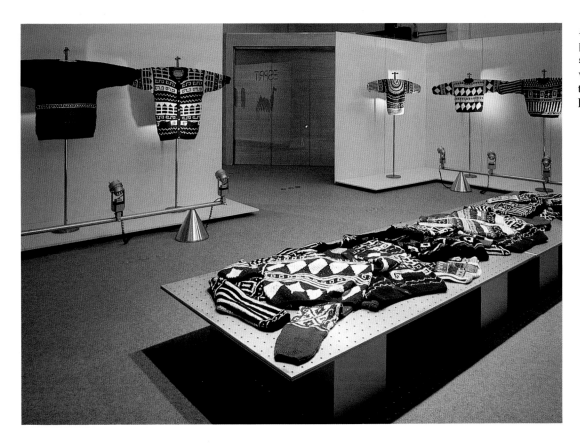

A simple setting, sharply lit, has no props to distract. The focus here is strictly on the merchandise. A wholesale showroom presentation by the Esprit San Francisco, Ca., Display Team.

Cookware customers can't miss the Calphalon corner at Emporium-Capwell in San Francisco, Ca. Tom Azzarello, V.P. of Visual Merchandising, uses a photo blow-up of the stock and makes a graphic point with the line's logo. An added service: the actual stock is pre-wrapped and ready to go.

Designer Thella Hall uses merchandise alone to set the tone in these Mother's Day cases at ZCMI.

The merchandise sets the mood and describes the lifestyle in this display case at ZCMI, Salt Lake City, Utah. Mike Stephens, Director, Thella Hall, Designer.

Designer Tim Davis shows shoppers
how to set a pretty spring table.
Also at ZCMI, Salt Lake City, Utah.

Bright ideas come up from the basement. ZCMI window by Sheldon Trimble, Supervisor/Designer.

CHAPTER V

Wit: Dangerous but Delicious

A common complaint now: "Stores are all alike these days. Same merchandise at the same price. If you didn't keep your wits about you, you'd forget which one you were shopping in."

On the next few pages you'll see the work of display designers who make stores seem unique by using their own true wit.

Funk & Wagnalls defines wit this way "... a keen perception of the incongruous or ludicrous ... a happy expression of unexpected or amusing analogies." So wit can be more than comedy: thought-provoking, surprising, attention-getting. And we all know it can be dangerous. A joke which is hilarious to one audience may seem dumb to another, or down-right insulting to a third. If a display scene approaches the outrageous, you can be sure customer complaints will continue until you change that window. Humor is such a touchy area that the late Morris Rosenblum, Creative Director of Macy's, established the "Groan Test." Ideas in this category were presented to colleagues, and if anybody groaned, the project was put on hold.

Still it's worth the risk. If you can make people stop, smile, act surprised or even shocked, you've made your establishment, at least for the moment, different from all others.

You'll find a wide range of wit on the ensuing pages — sharp-edged, gentle, ironic or cheerfully comical. And there's one popular trend to note. Professors of Visual Merchandising call it "elevating the mundane," using everyday objects in unexpected ways. This perfectly illustrates the dictionary's definition of wit: a keen perception of the incongruous.

Is this a struggle for women's rights? And what is that tea pot doing there? A stop-and-wonder window by Angela Patterson, Vice President, Director of Store Planning and Design, and Richard Currier, Director of Visual Presentation, for Bergdorf Goodman, New York.

Angela Patterson and Richard Currier use old black and white prints behind new short skirts. Can a glimpse of stocking still be shocking? For Bergdorf Goodman.

In this sculpture foam core figures reflect customer profiles at Johnston & Murphy, West Side Pavillion, Los Angeles, Ca. by R. Kennington Spikes, Visual Merchandising Manager.

Sculpture in the classic style serves as mannequins at Bergdorf Goodman, New York. By Angela Patterson and Richard Currier.

An upwardly mobile idea springs up at the Esprit Superstore in Los Angeles, Ca. By the Esprit Los Angeles Display Team.

Ordinary brown paper makes a surprising wrapper for this dashing sportswear. By English menswear designer Paul Smith for Paul Smith Inc., New York. Photography by Alex Emil.

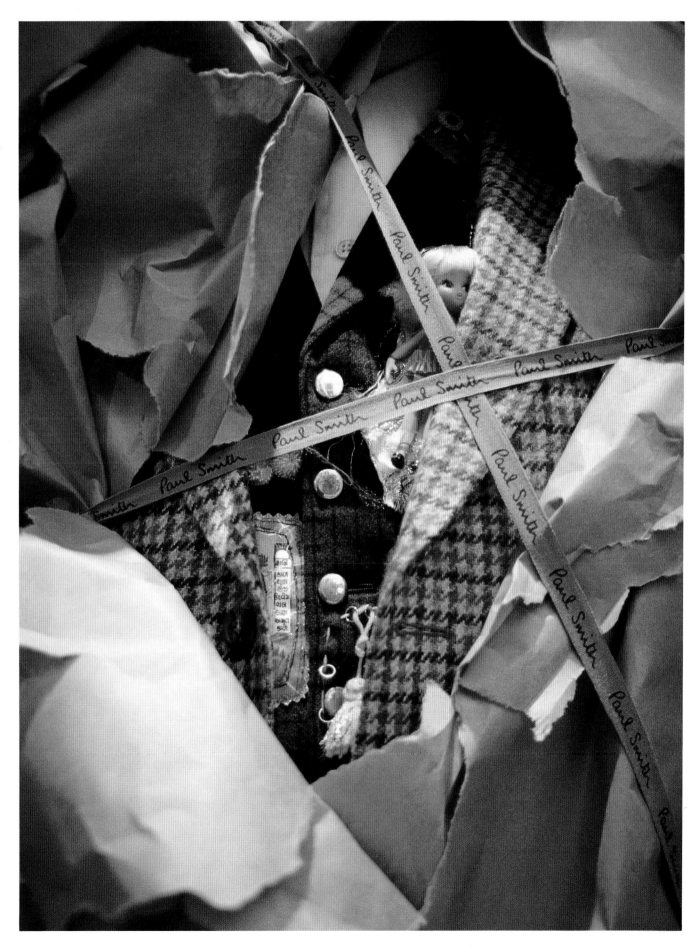

Danuta Ryder creates a wheat field at harvest time (and gives her mannequins wheat hair) to display clothes from a company named Workers for Freedom. Henri Bendel.

What could be more mundane than a pile of beer cans? Perch an incongruously elegant Baccarat eagle on top and you've built a witty window. Robert J. Mahoney, Display Director of Gump's in San Francisco, Ca., featured it during a convention of beverage makers.

Take me to the Taj Mahal. Danuta Ryder of Henri Bendel uses bright veils to create an Indian fantasy.

Whimsical wool. Everyday balls of yarn become beautiful sweaters in this window by Danuta Ryder for Henri Bendel, New York.

Fantasies in ivy. Witty lingerie display by Bendel's Danuta Ryder.

Everyday aluminum foil becomes a fantasy back-drop for evening wear. By Danuta Ryder for Henri Bendel.

Even blasé New Yorkers will stop to look at a three-legged lady. Danuta Ryder, Visual Planning Director of Henri Bendel on 57th Street, designs a hosiery window that also emphasizes store identity: the mannequin's dress is made of Bendel wrapping paper.

A ladder is just a ladder until you paint it red and put it in a fashion window. Then it becomes a stairway to success. By Danuta Ryder for Henri Bendel.

Danuta Ryder's people-pleasing puppet theatre, in the window of Henri Bendel.

A visual play on words: Henri Bendel's Danuta Ryder uses crunched seamless paper to form head-dresses. Can you guess the fashion company's name? Krunch.

When mystical rock crystals were the rage, they popped up at ZCMI, Salt Lake City, Utah. Ken Diamond, designer.

Carmen Miranda would love this beauty pageant. Henri Bendel's Danuta Ryder calls the swimwear window "frutti-ful."

CHAPTER **VI**

Graphics, GRAPHICS, *graphics,* graphics

Here's a vital trend in visual selling, one that's worth careful watching. Designs presented in the previous chapters bear a definite relationship to major schools of fine art. You'll recall seeing Classical works from Bergdorf Goodman, Impressionism from Henri Bendel, Surrealism from ZCMI, as well as Minimalist, Modern and especially Romanticist styles.

This chapter celebrates a genre that's the opposite of whimsy and romance — graphics. This can encompass Realism, or design so pure it's almost Abstract; you'll see no fantasy sequences among these photographs.

One refreshing trend emerges: words. The creators of most of these sets have rediscovered the value of language as a selling tool, using lettering — delicate or dramatic — to focus attention on their merchandising messages. For that, congratulations to the artists represented here. May others follow your lead.

Management and creative teams might also consider this: perhaps audiences are ready to face and embrace reality.

Cut-out shapes make a dramatic graphic background in this ZCMI window by Thella Hall.

Supervisor/Designer Anne Cook of ZCMI plays up the beauty of brass with glowing graphics.

Sharply defined design, in almost abstract black and white, and equally bold graphics. A ZCMI display by designer Tim Taylor.

Life is bright in Salt Lake City, according to ZCMI designer Sue Hansen's use of colors, clothes and words.

Elegant graphics add to the impact of this fashion window by ZCMI Supervisor/Designer Diane Call.

This watch window makes use of bold graphics and interesting props, like rusty pipes and bricks. Also by Roy Lauritzen for ZCMI, Salt Lake City.

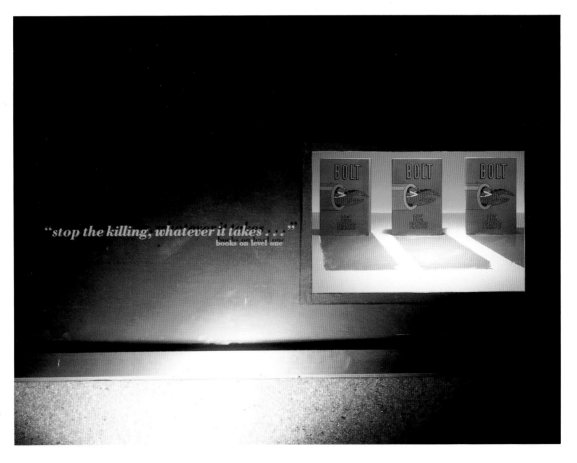

Designer Sue Hansen uses subtle italics to contrast with the dramatic quote in a book-selling window at ZCMI.

Graphics done with copy and a pleasant pun are a strong focal point for this gift display at ZCMI, Salt Lake City, Utah. Mike Stephens, Director of Visual Merchandising; Roy Lauritzen, Designer.

The background may be idealized, but strong graphics announce a down-to-earth event at ZCMI, Salt Lake City. Mike Stephens, Director of Visual Merchandising; Ken Diamond, designer.

A smile-provoking set-up by designer Ken Diamond. For ZCMI.

Designer Thella Hall of ZCMI, Salt Lake City, uses simple graphics and a wooden bar to simulate a cruise ship.

Bloomingdale's New York uses loose and easy lettering to match the happy-days feeling of their California collection. By Joe Feczko, Vice President of Visual Presentation, and Myk Fisher, Ready-to-Wear Designer, Visual Presentation.

Beachin' it at Bloomies. By Richard Knapple for Bloomingdale's New York.

Riding an abstract surf at Bloomingdale's New York. By Joe Feczko, Vice President of Visual Presentation, and Myk Fisher, Ready-to-Wear Designer, Visual Presentation.

Homer Sharp goes over our heads with graphics announcing Hermes at Marshall Field's.

What is the name of the fragrance launched at Marshall Field's of Chicago, Ill.? Homer G. Sharp, Vice President, Design, uses every device to tell us it's Red, including ingenious graphics on the floor.

By Homer G. Sharp for Marshall Field's, Chicago, Ill.

An elegant mannequin surrounded by soft graphics. G for Glamour and G for Galanos. Angela Patterson, Vice President, Director of Store Planning and Design and Richard Currier, Director of Visual Presentation. Bergdorf Goodman, New York.

The Esprit logo, with a blown-up photo of products, makes it graphically clear that this is an accessories-only kiosk. A San Francisco, Ca., wholesale showroom by Esprit's San Francisco Display Team.

Designer Roy Lauritzen knows how to focus attention on his store's name.

Black and white and a splash of bright. Sophisticated setting and graphics to match at ZCMI. Mike Stephens, Director; Roy Lauritzen, designer.

CHAPTER VII

Great Spaces: an Endangered Species?

Around the turn of the century, when the United States was still young and growing, expansion and optimism were the operative words. The country seemed endlessly open. There was room enough for everything and anything, including stores built on a grand scale. So grand, indeed, that newspapers of that era referred to them as "palaces of trade."

Sadly, only a few of these palatial emporia have survived unspoiled. As we know, many of those great buildings were converted to other uses when lavish size made them economically unviable as stores. Others have been modernized: ceilings lowered, atria filled in, balconies removed.

These next examples show not only glorious old stores, but how some creative store planners are discovering new, less costly ways to utilize big areas, or to give the illusion of airy, open vistas where they don't actually exist.

This chapter is devoted to those remaining palace-sized stores, to new sorts of space use, and especially to those designers who know how to make the most of vast dimensions.

Perhaps these photographs will serve as inspiration to visual-design professionals who find that wide open aisles and towering heights are sometimes more of a problem than a pleasure.

Ornate urns complement the classic capitals of towering columns. By Homer G. Sharp for Marshall Field's, Chicago.

One reason why fine fragrance houses love to launch a new scent at Marshall Field's in Chicago, Ill.: the store's great space makes for great drama. For Nina by Nina Ricci, Design Vice President Homer G. Sharp took advantage of height (to create lavish swags) and width (to roll out a luxurious, graphically focused carpet). This atrium, originally called a "light dome," is 14 stories tall; the wonderfully wide aisle is 75 feet long.

Marshall Field's, Chicago, Ill.
Design by Homer G. Sharp.

This photograph gives a sense of the impressive height of Marshall Field's atrium. Homer Sharp, Design Vice President.

Another advantage of height: mannequins can stand comfortably far above counter tops. By Homer Sharp for Marshall Field's.

Romantic lighting draws attention to mannequins and merchandise and creates pools of interest within big space. By Homer Sharp for Marshall Field's.

Here's a 21st century approach to design for big space — simple, almost sassy, industrial and dramatic. Don't miss the important graphics and total focus on merchandise. A showroom in San Francisco, Ca. by the Esprit San Francisco Visual Display team.

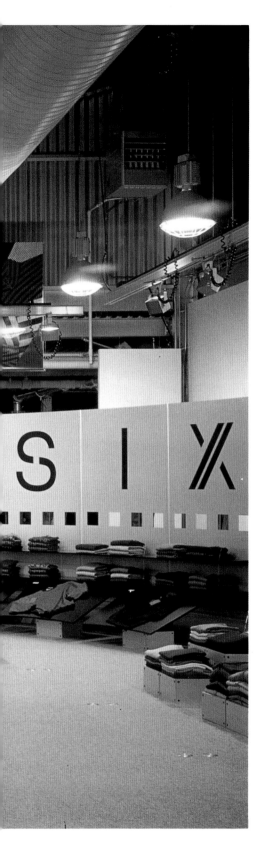

An innovative idea by William F. Short, Director of Visual Merchandising for Parisian. He creates the illusion of super-spaciousness by placing a barreled ceiling of "sky" over an escalator well at the Parisian at Forest Fair Mall, Cincinnati, Ohio.

CHAPTER **VIII**

Small Space: a Special Sort of Charm

This book wouldn't be complete without a little section in praise of the petite. The pictures in this and the preceding chapter are proof positive of the two most important tenets in a visual merchandiser's career — *Positive thinking* is powerful enough to overcome almost any problem, and *Fine design* can bring excitement to tiny corners as well as to cavernous spaces. If thousands of square feet can seem daunting, a meager few can be viewed that way too.

Chapter VII honors creators who make big areas breathtaking. These pages applaud the wit and wisdom of those who make little spaces loveable.

Here's to marvelous miniatures and to those who make them so mesmerizing.

Little kids, little window, lots of heart. An Esprit Valentine design by Tim Davis, ZCMI, Salt Lake City.

Slowly falling drops of water create a mystical mood for lavender jade jewelry at Gump's San Francisco. By James H. Stearns, Assistant Display Director. Mechanism and installation by Kent Bond.

For Gump's, a Baccarat crystal butterfly is accompanied by an Hakata doll. "Madame Butterfly" window by Display Director Robert J. Mahoney. Execution by Kent Bond. Signage by Jean Dolmans.

A rug sale takes to the air with flying carpets that are actually dollhouse sized. Robert J. Mahoney, SVM, Display Director of Gump's in San Francisco, Ca., adds flair to an otherwise ordinary annual event.

An interesting mix of textures: tin flowers in a precious Oriental cache pot. A nifty little window by Robert Mahoney for Gump's.

A common retail commodity, Paul Revere bowls, become irresistible in this little window. Note the interesting graphics and the patriotic red, white and blue mylar tape. By Robert J. Mahoney for Gump's San Francisco. Photography by Troy Staten.

A small table becomes a perfect showplace for Southwest art and accessories. By Joanna Seitz and Robert Whittingham for J. Seitz & Co., New Preston, Conn.

Pale, gleaming floors and glowing glass give a sense of space and open horizons at Bonwit Teller in Troy, Mich. by F.C. Calise, Vice President, Visual Merchandising and Design.

Even the merchandise is wee and witty in this small window at Bergdorf Goodman, New York. Necklace is made of miniature ceramic plates. Angela Patterson, Vice President, Director of Store Planning and Design and Richard Currier, Director of Visual Presentation.

Delightful is the word for this little display case by designer Thella Hall of ZCMI, Salt Lake City, Utah.

Robert Mahoney uses specially designed graphics to spell out the name of a new dinner pattern. Letter colors match the plates' borders. Lettering design by Jean Dolmans. Troy Staten, photographer.

Small window at Gump's, giant asparagus. A clever way to show customers what this serving set is used for. Incidentally, Robert Mahoney uses small spaces by choice, not by necessity. Gump's full-sized windows are masked or painted to allow focus on these charming miniature sets. Photograph by Troy Staten.

CHAPTER IX

Christmas Glories and Other Gala Events

The Christmas holidays bring pleasure and profit to customers and merchants alike. Shoppers have the pleasure of seeing fascinating holiday windows and fabulous decorations. And they profit from new ideas presented at fashion seminars and how-to demonstrations. Merchants profit from added sales, of course, along with the good publicity that results from excellent displays and well-staged events.

Visual design teams? They do the brunt of the hard work, but they also reap the rewards of a job well done: all the delighted reactions from the day windows open until the last bee light comes down.

A little frivolous frosting is often the fun in American retailing. Let's hope there will always be enough time, talent and funding for grand gestures like the ones highlighted here.

"Fabled Lands" was the name of this month-long promotion of exotic import merchandise at Gump's San Francisco, Ca. Robert J. Mahoney, Display Director; James H. Stearns, Assistant Display Director. Paintings by James Stearns, Bruce Henderson and Kent Bond. Accessorized by Scott Butcher. Troy Staten, photographer.

Gump's adopts the San Francisco S.P.C

Public service combined with pleasure: "Gump's Adopts the San Francisco SPCA" Christmas windows encouraged animal adoption, charmed the public and won national publicity. Robert J. Mahoney, SVM, Display Director; James H. Stearns, Assistant Display Director; Bruce Henderson, Staff Artist. Photography by Troy Staten.

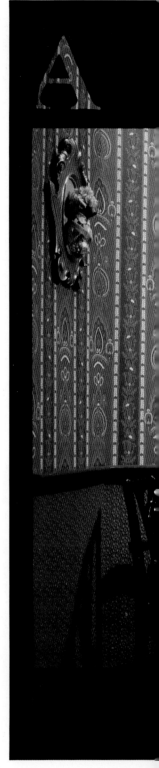

A variation on a successful theme appears in another set of windows, as adoptive animals go on a Christmas cruise. Robert Mahoney, James Stearns and Bruce Henderson. Animal costumes by Larry Bianchi and John Gilkerson.

Trees, lights, dolls and delights for the merry main aisle of Marshall Field's in Chicago, Ill. Homer G. Sharp, SVM, Vice President, Design.

Traditional Christmas glory: the great tree that fills the atrium at the Emporium-Capwell in San Francisco, Ca. Thomas Azzarello, Vice President of Visual Merchandising.

Homer G. Sharp, Vice President, Design, uses new and retro fashion, with a street scene backdrop, as part of a fragrance event at Marshall Field's, Chicago, Ill.

A magazine tie-in promotion offers color, snap and smart graphics. By designer Roy Lauritzen for ZCMI.

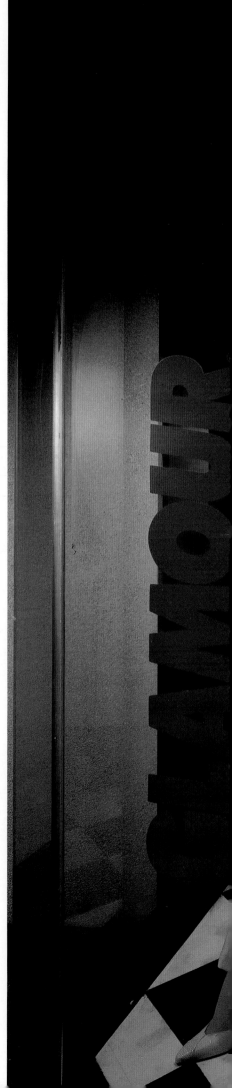

OLOR & ST

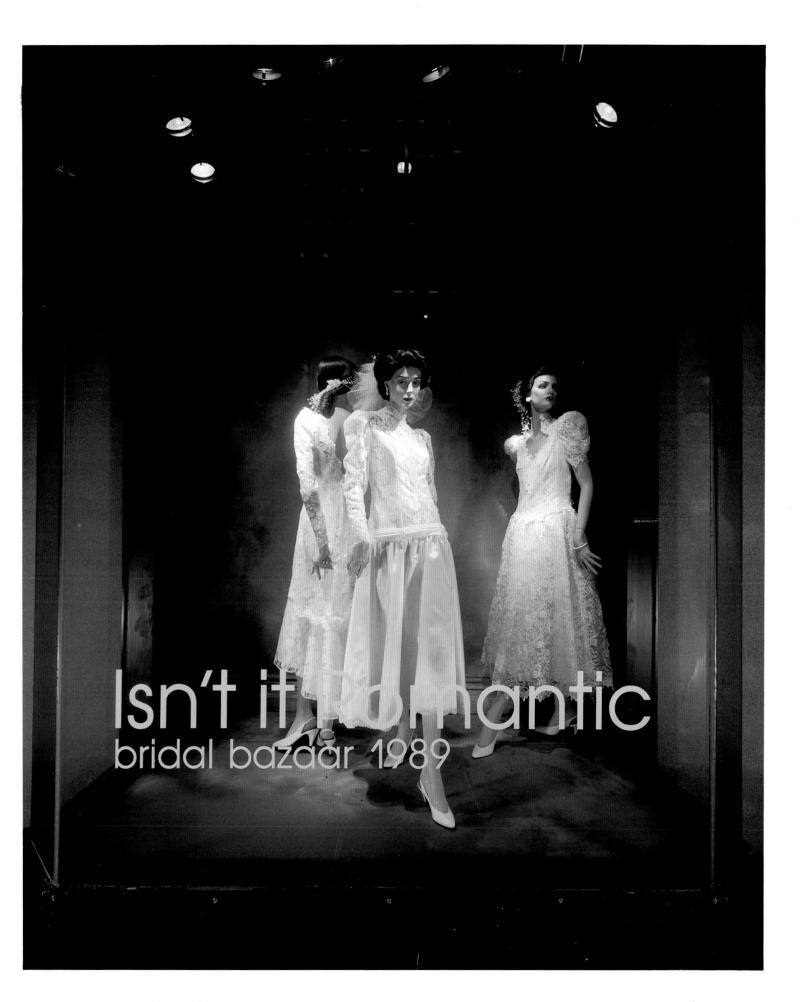

Brides are traditional and beautiful at ZCMI. Diane Call and Ken Diamond designed these windows to promote an annual event in Salt Lake City.

Isn't it Romantic
bridal bazaar 1989

The Easter Bunny still comes to Salt Lake City, Utah. Mike Stephens, Director, Visual Merchandising. Diane Call, Supervisor/Designer. Diane Pierce, Designer. ZCMI.

Another Mother's Day window at ZCMI. Roy Lauritzen, designer.

Photos in this Mother's Day window are store employees and their real life Moms. By Anne Cook, Supervisor/Designer for ZCMI.